THE TRUE WIRELESS

by Nikola Tesla

Electrical Experimenter, May 1919

For information regarding special discounts for bulk purchases, please contact BN Publishing at info@bnpublishing.net

www.bnpublishing.net

Printed in the U.S.A.

THE TRUE WIRELESS

by Nikola Tesla
Electrical Experimenter, May 1919

Ever since the announcement of Maxwell's electro-magnetic theory scientific investigators all the world over had been bent on its experimental verification. They were convinced that it would be done and lived in an atmosphere of eager expectancy, unusually favorable to the reception of any evidence to this end. No wonder then that the publication of Dr. Heinrich Hertz's results caused a thrill as had scarcely ever been experienced before. At that time I was in the midst of pressing work in connection with the commercial introduction of my system of power transmission, but, nevertheless, caught the fire of enthusiasm and fairly burned with desire to behold the miracle with my own eyes. Accordingly, as soon as I had freed myself of these imperative duties and resumed research work in my laboratory on Grand Street, New York, I began, parallel with high frequency alternators, the construction of several forms of apparatus with the object of exploring the field opened up by Dr. Hertz. Recognizing the limitations of the devices he had employed, I concentrated my attention on the production of a powerful induction coil but made no notable progress until a happy inspiration led me to the invention of the oscillation transformer. In the latter part of 1891 I was already so far advanced in the development of this new principle that I had at my disposal means vastly superior to those of the German physicist. All my previous efforts with Rhumkorf coils had left me unconvinced, and in order to settle my doubts I went over the whole ground once more, very carefully, with these improved appliances. Similar phenomena were noted, greatly magnified in intensity, but they were susceptible of a different and more plausible explanation. I considered this so important that in 1892 I went to Bonn, Germany, to confer with Dr. Hertz in regard to my observations. He seemed disappointed to such a degree that I regretted my trip and parted from him sorrowfully. During the succeeding years I made numerous experiments with the same object, but the results were uniformly negative. In 1900, however, after I had evolved a wireless transmitter which enabled me to obtain electro-magnetic activities of many millions of horse-power, I made a last desperate attempt to prove that the disturbances emanating from the oscillator were ether vibrations akin to those of light, but met again with utter failure. For more than eighteen years I have been reading treatises, reports of scientific transactions, and articles on Hertz-wave telegraphy, to keep myself informed, but they have always imprest me like works of fiction.

The history of science shows that theories are perishable. With every new truth that is revealed we get a better understanding of Nature and our conceptions and views are modified. Dr. Hertz did not discover a new principle. He merely gave material support to hypothesis which had been long ago formulated. It was a perfectly well-established fact that a circuit, traversed by a periodic current, emitted some kind of space waves, but we were in ignorance as to their character. He apparently gave an experimental proof that they were transversal vibrations in the ether. Most people look upon this as his great accomplishment. To my mind it seems that his immortal merit was not so much in this as in the focusing of the investigators' attention on the processes taking place in the ambient medium. The Hertz-wave theory, by its fascinating hold on the imagination, has stifled creative effort in the wireless art and retarded it for twenty-five years. But, on the other hand, it is impossible to over-estimate the beneficial effects of the powerful stimulus it has given in many directions.

As regards signaling without wires, the application of these radiations for the purpose was quite obvious. When Dr. Hertz was asked whether such a system would be of practical value, he did not think so, and he was correct in his forecast. The best that might have been expected was a method of communication similar to the heliographic and subject to the same or even greater limitations.

In the spring of 1891 I gave my demonstrations with a high frequency machine before the American Institute of Electrical Engineers at Columbia College, which laid the foundation to a new and far more promising departure. Altho the laws of electrical resonance were well known at that time and my lamented friend, Dr. John Hopkinson, had even indicated their specific application to an alternator in the Proceedings of the Institute of Electrical Engineers, London, Nov.13, 1889, nothing had been done towards the practical use of this knowledge and it is probable that those experiments of mine were the first public exhibition with resonant circuits, more particularly of high frequency. While the spontaneous success of my lecture was due to spectacular features, its chief import was in showing that all kinds of devices could be operated thru a single wire without return. This was the initial step in the evolution of my wireless system. The idea presented itself to me that it might be possible, under observance of proper conditions of resonance, to transmit electric energy thru the earth, thus dispensing with all artificial conductors. Anyone who might wish to examine impartially the merit of that early suggestion must not view it in the light of present day science. I only need to say that as late as 1893, when I had prepared an elaborate chapter on my wireless system, dwelling on its various instrumentalities and future prospects, Mr. Joseph Wetzler and other friends of mine emphatically protested against its publication on the ground that such idle and far-fetched speculations would injure me in the opinion of conservative business men. So it came that only a small part of what I had intended to say was embodied in my address of that year before the Franklin Institute and National Electric Light Association under the chapter "On Electrical Resonance." This little salvage from the wreck has earned me the title of "Father of the Wireless" from many well-disposed fellow workers, rather than the invention of scores of appliances which have brought wireless transmission within the reach of every young amateur and which, in a time not distant, will lead to undertakings overshadowing in magnitude and importance all past achievements of the engineer.

The popular impression is that my wireless work was begun in 1893, but as a matter of fact I spent the two preceding years in investigations, employing forms of apparatus, some of which were almost like those of today. It was clear to me from the very start that the successful consummation could only be brought about by a number of radical improvements. Suitable high frequency generators and electrical oscillators had first to be produced. The energy of these had to be transformed in effective transmitters and collected at a distance in proper receivers. Such a system would be manifestly circumscribed in its usefulness if all extraneous interference were not prevented and exclusiveness secured. In time, however, I recognized that devices of this kind, to be most effective and efficient, should be designed with due regard to the physical properties of this planet and the electrical conditions obtaining on the same. I will briefly touch upon the salient advances as they were made in the gradual development of the system.

The high frequency alternator employed in my first demonstrations is illustrated in Fig. 1. It comprised a field ring, with 384 pole projections and a disc armature with coils wound in one single layer which were connected in various ways according to requirements. It was an excellent machine for experimental purposes, furnishing sinusoidal currents of from 10,000 to 20,000 cycles per second. The output was comparatively large, due to the fact that as much as 30 amperes per square millimeter could be past thru the coils without injury.

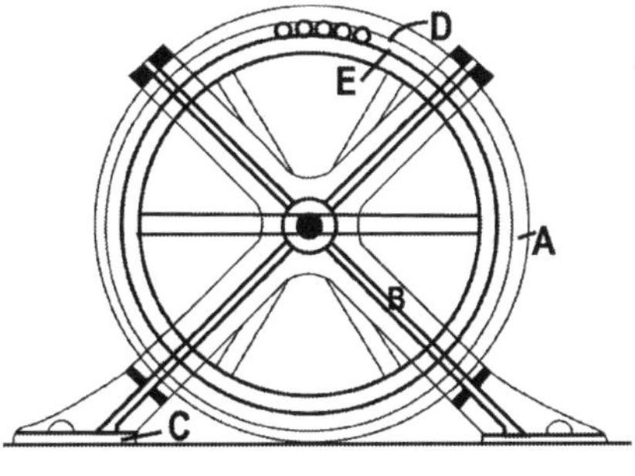

Fig. 1. Alternator of 10.000 Cycles p.s., Capacity 10K.W.. Which Was Employed by Tesla in His First Demonstrations of High Frequency Phenomena Before the American Institute of Electrical Engineers at Columbia College, May 20, 1891.

The diagram in Fig. 2 shows the circuit arrangements as

used in my lecture. Resonant conditions were maintained by means of a condenser subdivided into small sections, the finer adjustments being effected by a movable iron core within an inductance coil. Loosely linked with the latter was a high tension secondary which was tuned to the primary.

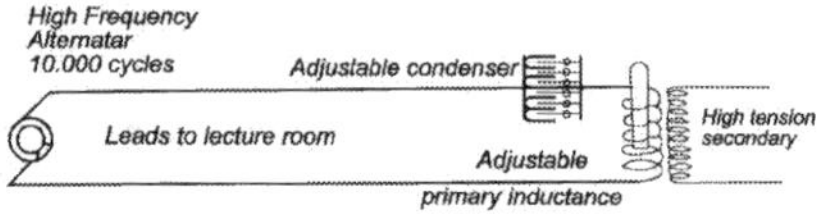

Fig. 2. Diagram Illustrating the Circuit Connections and Tuning Devices Employed by Tesla In His Experimental Demonstrations Before the American Institute of Electrical Engineers With the High Frequency Alternator Shown in Fig. 1.

The operation of devices thru a single wire without return was puzzling at first because of its novelty, but can be readily explained by suitable analogs. For this purpose reference is made to Figs. 3 and 4.

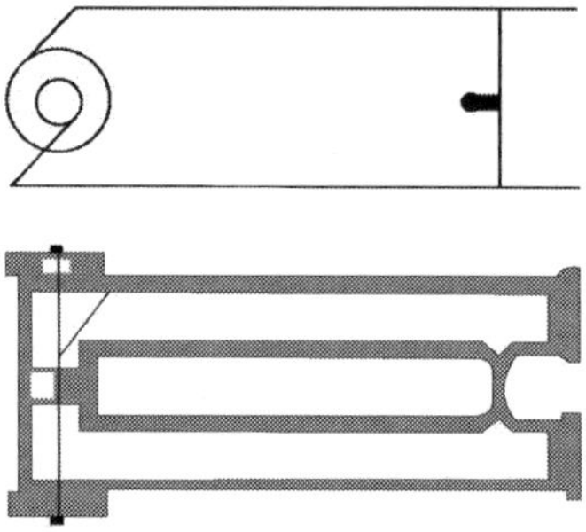

Fig. 3. Electric Transmission Thru Two Wires and Hydraulic Analog.

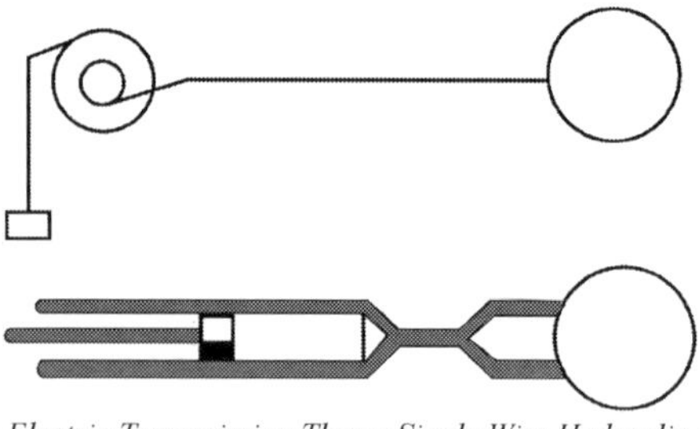

Fig. 4. Electric Transmission Thru a Single Wire Hydraulic Analog.

In the former the low resistance electrical conductors are represented by pipes of large cross section, the alternator by an oscillating piston and the filament of an incandescent lamp by a minute channel connecting the pipes. It will be clear from a glance at the diagram that very slight excursions of the piston would cause the fluid to rush with high velocity thru the small channel and that virtually all the energy of movement would be transformed into heat by friction, similarly to that of the electric current in the lamp filament.

The second diagram will now be self-explanatory. Corresponding to the terminal capacity of the electric system an elastic reservoir is employed which dispenses with the necessity of a return pipe. As the piston oscillates the bag expands and contracts, and the fluid is made to surge thru the restricted passage with great speed, this resulting in the generation of heat as in the incandescent lamp. Theoretically considered, the efficiency of conversion of energy should be the same in both cases.

Granted, then, that an economic system of power transmission thru a single wire is practicable, the question arises how to collect the energy in the receivers. With this object attention is called to Fig. 5, in which a conductor is shown excited by an oscillator joined to it at one end. Evidently, as the periodic impulses pass thru the wire, differences of potential will be created along the same as well as at right angles to it in the surrounding medium and either of these may be usefully applied. Thus at a, a circuit comprising an inductance and capacity is resonantly excited in the transverse, and at b, in the longitudinal sense. At c, energy is collected in a circuit parallel to the conductor but not in contact with it, and again at d, in a circuit which is partly sunk into the conductor and may be, or not, electrically connected to the same. *It is important to keep these typical dispositions in mind, for however the distant actions of the oscillator might be modified thru the immense extent of the globe the principles involved are the same.*

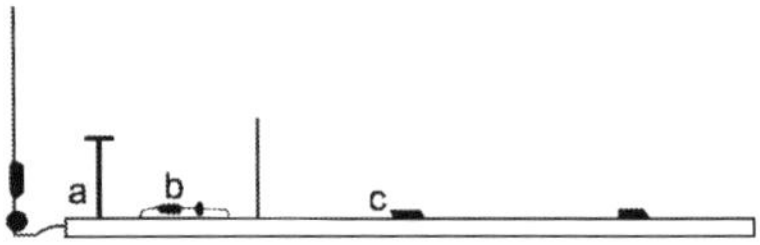

Fig. 5. Illustrating Typical Arrangements for Collecting Energy In a System of Transmission Thru a Single Wire.

Consider now the effect of such a conductor of vast dimensions on a circuit exciting it. The upper diagram of Fig. 6 illustrates a familiar oscillating system comprising a straight rod of self-inductance 2L with small terminal capacities cc and a node in the center. In the lower diagram of the figure a large capacity C is attached to the rod at one end with the result of shifting the node to the right, thru a distance corresponding to self-inductance X. As both parts of the system on either side of the node vibrate at the same rate, we have evidently, $(L+X)c = (L-X)C$ from which $X = L(C-c/C+c)$. When the capacity C becomes commensurate to that of the earth, X approximates L, in other words, the node is close to the ground connection. *The exact determination of its position is very important in the calculation of certain terrestrial electrical and geodetic data* and I have devised special means with this purpose in view.

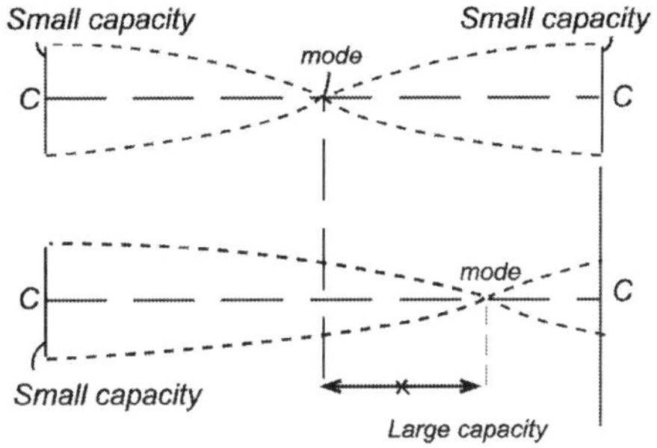

Fig. 6. Diagram Elucidating Effect of Large Capacity on One End.

My original plan of transmitting energy without wires is shown in the upper diagram of Fig. 7, while the lower one illustrates its mechanical analog, first published in my article in the Century Magazine of June, 1900. An alternator, preferably of high tension, has one of its terminals connected to the ground and the other to an elevated capacity and impresses its oscillations upon the earth. At a distant point a receiving circuit, likewise connected to ground and to an elevated capacity, collects some of the energy and actuates a suitable device. I suggested a multiplication of such units in order to intensify the effects, an idea which may yet prove valuable. In the analog two tuning forks are provided, one at the sending and the other at the receiving station, each having attached to its lower prong a piston fitting in a cylinder. The two cylinders communicate with a large elastic reservoir filled with an incompressible fluid. The vibrations transmitted to either of the tuning forks excite them by resonance and, thru electrical contacts or otherwise, bring about the desired result. This, I may say, was not a mere mechanical illustration, but a simple representation of my apparatus for submarine signaling, perfected by me in 1892, but not appreciated at that time, altho more efficient than the instruments now in use.

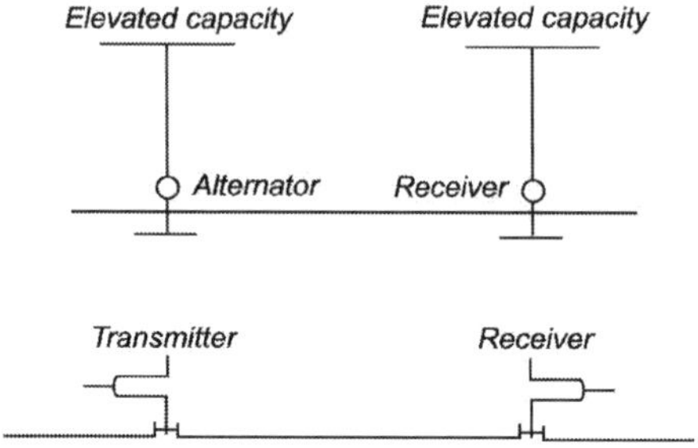

Fig. 7. Transmission of Electrical Energy Thru the Earth as Illustrated in Tesla's Lectures Before the Franklin Institute and Electric Light Association in February and March, 1893, and Mechanical Analog of the Same.

The electric diagram in Fig. 7, which was reproduced from my lecture, was meant only for the exposition of the principle. The arrangement, as I described it in detail, is shown in Fig. 8. In this case an alternator energizes the primary of a transformer, the high tension secondary of which is connected to the ground and an elevated capacity and tuned to the imprest oscillations. The receiving circuit consists of an inductance connected to the ground and to an elevated terminal without break and is resonantly responsive to the transmitted oscillations. A specific form of receiving device was not mentioned, but I had in mind to transform the received currents and thus make their volume and tension suitable for any purpose. This, in substance, is the system of today and I am not aware of a singe authenticated instance of successful transmission at considerable distance by different instrumentalities. It might, perhaps, not be clear to those who have perused my first description of these improvements that, besides making known new and efficient types of apparatus, I gave to the world a wireless system of potentialities far beyond anything before conceived. I made explicit and repeated statements that I contemplated transmission, absolutely unlimited as to terrestrial distance and amount of energy. But, altho I have overcome all obstacles which seemed in the beginning unsurmountable and found elegant solutions of all the problems which confronted me, yet, even at this very day, the majority of experts are still blind to the possibilities which are within easy attainment.

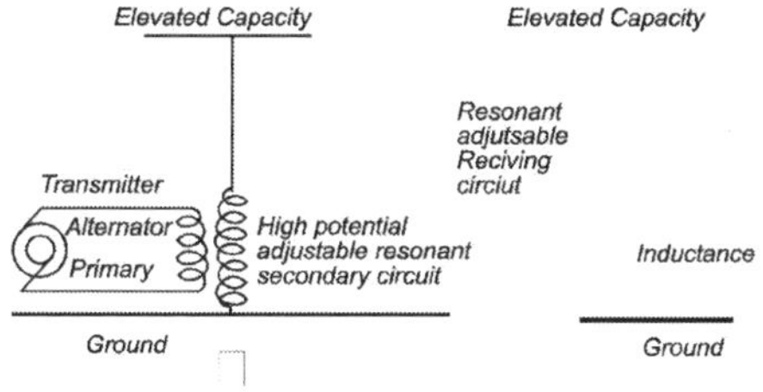

Fig. 8. Tesla's System of Wireless Transmission Thru the Earth as Actually Exposed In His Lectures Before the Franklin Institute and Electric Light Association in February and March, 1893.

My confidence that a signal could be easily flashed around the globe was strengthened thru the discovery of the "rotating brush," a wonderful phenomenon which I have fully described in my address before the Institution of Electrical Engineers, London, in 1892 , and which is illustrated in Fig. 9. This is undoubtedly the most delicate wireless detector known, but for a long time it was hard to produce and to maintain in the sensitive state. These difficulties do not exist now and I am looking to valuable applications of this device, particularly in connection with the high-speed photographic method, which I suggested, in wireless, as well as in wire, transmission.

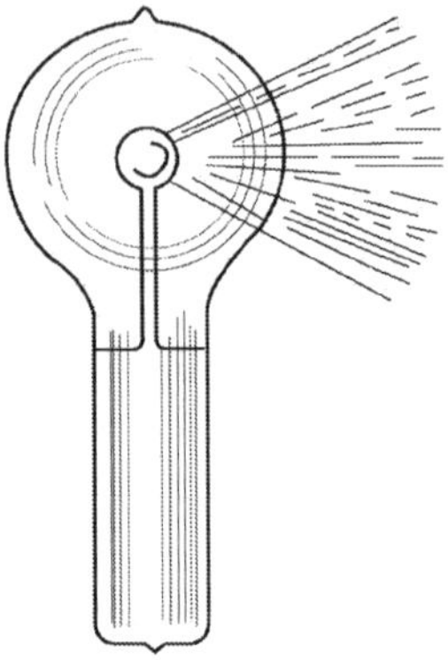

Fig. 9. The Forerunner of the Audion-the Most Sensitive Wireless Detector Known, as Described by Tesla In His Lecture Before the Institution of Electrical Engineers, London, February, 1892.

Possibly the most important advances during the following three or four years were my system of concatenated tuned circuits and methods of regulation, now universally adopted. The intimate bearing of these inventions on the development of the wireless art will appear from Fig. 10, which illustrates an arrangement described in my U.S. Patent No. 568,178 of September 22, 1896, and corresponding dispositions of wireless apparatus. The captions of the individual diagrams are thought sufficiently explicit to dispense with further comment. I will merely remark that in this early record, in addition to indicating how any number of resonant circuits may be linked and regulated, I have shown the advantage of the proper timing of primary impulses and use of harmonics. In a farcical wireless suit in London, some engineers, reckless of their reputation, have claimed that my circuits were not at all attuned; in fact they asserted that I had looked upon resonance as a sort of wild and untamable beast!

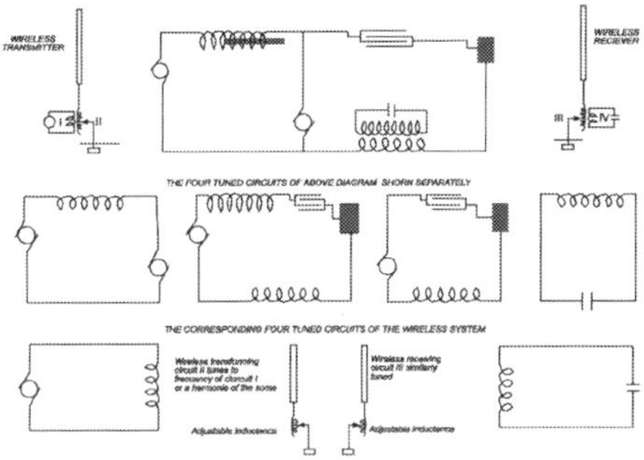

Fig. 10. Tesla's System of Concatenated Tuned Circuits Shown and Described In U. S. Patent No. 568,178 of September 22, 1896, and Corresponding Arrangements in Wireless Transmission.

It will be of interest to compare my system as first described in a Belgian patent of 1897 with the Hertz-wave system of that period. The significant differences between them will be observed at a glance. The first enables us to transmit economically energy to any distance and is of inestimable value; the latter is capable of a radius of only a few miles and is worthless. In the first there are no spark-gaps and the actions are enormously magnified by resonance. In both transmitter and receiver the currents are transformed and rendered more effective and suitable for the operation of any desired device. Properly constructed, my system is safe against static and other interference and the amount of energy which may be transmitted is *billions of times greater* than with the Hertzian which has none of these virtues, has never been used successfully and of which no trace can be found at present.

A well-advertised expert gave out a statement in 1899 that my apparatus did not work and that it would take 200 years before a message would be flashed across the Atlantic and even accepted stolidly my congratulations on a supposed great feat.

But subsequent examination of the records showed that my devices were secretly used all the time and ever since I learned of this I have treated these Borgia-Medici methods with the contempt in which they are held by all fair-minded men. The wholesale appropriation of my inventions was, however, not always without a diverting side. As an example to the point I may mention my oscillation transformer operating with an air gap. This was in turn replaced by a carbon arc, quenched gap, an atmosphere of hydrogen, argon or helium, by a mechanical break with oppositely rotating members, a mercury interrupter or some kind of a vacuum bulb and by such *tours de force* as many new "systems" have been produced. I refer to this of course, without the slightest ill-feeling, let us advance by all means. But I cannot help thinking how much better it would have been if the ingenious men, who have originated these "systems," had invented something of their own instead of depending on me altogether.

Before 1900 two most valuable improvements were made. One of these was my individualized system with transmitters emitting a wave-complex and receivers comprising separate tuned elements cooperatively associated. The underlying principle can be explained in a few words. Suppose that there are n simple vibrations suitable for use in wireless transmission, the probability that any one tune will be struck by an extraneous disturbance is 1/n. There will then remain n-1 vibrations and the chance that one of these will be excited is 1/n-1 hence the probability that two tunes would be struck at the same time is 1/n(n-1). Similarly, for a combination of three the chance will be 1/n(n-1)(n-2) and so on. It will be readily seen that in this manner any desired degree of safety against the statics or other kind of disturbance can be attained provided the receiving apparatus is so designed that is operation is possible only thru the joint action of all the tuned elements. This was a difficult problem which I have successfully solved so that now *any desired number of simultaneous messages is practicable in the transmission thru the earth as well as thru artificial conductors.*

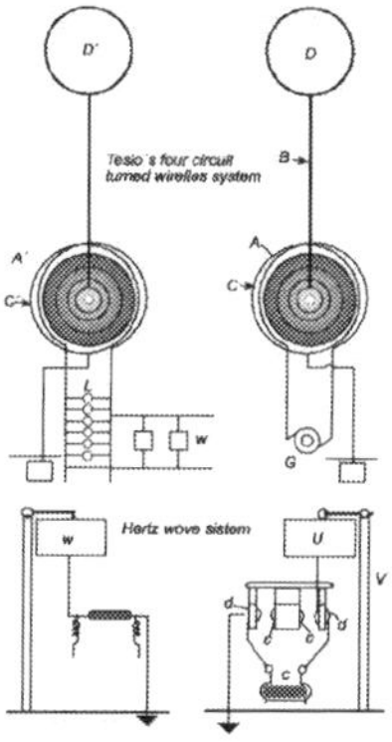

Fig. 11. Tesla's Four Circuit Tuned System Contrasted With the Contemporaneous Hertz-wave System.

The other invention, of still greater importance, is a peculiar oscillator enabling the transmission of energy without wires in any quantity that may ever be required for industrial use, to any distance, and with very high economy. It was the outcome of years of systematic study and investigation and wonders will be achieved by its means.

The prevailing misconception of the mechanism involved in the wireless transmission has been responsible for various unwarranted announcements which have misled the public and worked harm. By keeping steadily in mind that the transmission thru the earth is in every respect identical to that thru a straight wire, one will gain a clear understanding of the phenomena and will be able to judge correctly the merits of a new scheme. Without wishing to detract from the value of any plan that has been put forward I may say that they are devoid of novelty. So for instance in Fig. 12 arrangements of transmitting and receiving circuits are illustrated, which I have described in my U.S. Patent No. 613,809 of November 8, 1898 on a Method of and Apparatus for Controlling Mechanism of Moving Vessels or Vehicles, and which have been recently dished up as original discoveries. In other patents and technical publications I have suggested conductors in the ground as one of the obvious modifications indicated in Fig. 5.

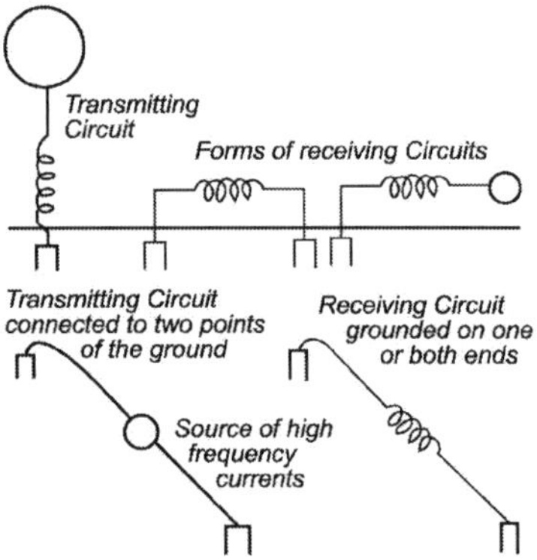

Fig. 12. Arrangements of Directive Circuits Described In Tesla's U. S. Patent No. 613,809 of November 8. 1898, on "Method of and Apparatus for Controlling Mechanism of Moving Vessels or Vehicles."

For the same reason the statics are still the bane of the wireless. There is about as much virtue in the remedies recently proposed as in hair restorers. *A small and compact apparatus has been produced which does away entirely with this trouble*, at least in plants suitably remodeled.

Nothing is more important in the present phase of development of the wireless art than to dispose of the dominating erroneous ideas. With this object I shall advance a few arguments based on my own observations *which prove that Hertz waves have little to do with the results obtained even at small distances.*

In Fig. 13 a transmitter is shown radiating space waves of considerable frequency. It is generally believed that these waves pass along the earth's surface and thus affect the receivers. I can hardly think of anything more improbable than this "gliding wave" theory and the conception of the "guided wireless" which are contrary to all laws of action and reaction. Why should these disturbances cling to a conductor where they are counteracted by induced currents, when they can propagate in all other directions unimpeded? The fact is that the radiations of the transmitter passing along the earth's surface are soon extinguished, the height of, the inactive zone indicated in the diagram, being some function of the wave length, the bulk of the waves traversing freely the atmosphere. Terrestrial phenomena which I have noted conclusively show that there is no *Heaviside layer*, or if it exists, it is of no effect. It certainly would be unfortunate if the human race were thus imprisoned and forever without power to reach out into the depths of space.

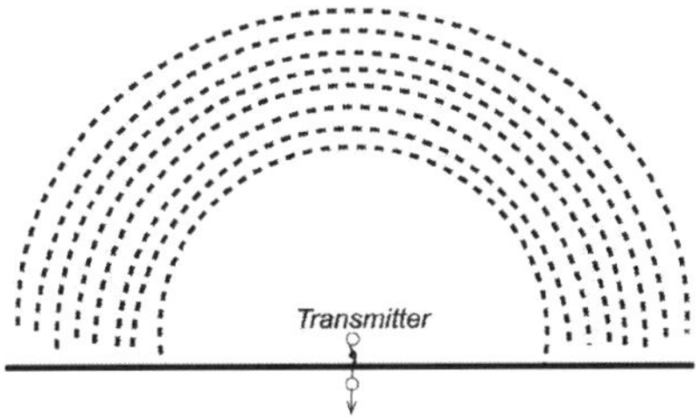

Fig. 13. Diagram Exposing the Fallacy of the Gilding Wave Theory as Propounded In Wireless Text Books.

The actions at a distance cannot be proportionate to the height of the antenna and the current in the same. I shall endeavor to make this clear by reference to diagram in Fig. 14. The elevated terminal charged to a high potential induces an equal and opposite charge in the earth and there are thus Q lines giving an average current I=4Qn which circulates locally and is useless except that it adds to the momentum. A relatively small number of lines q however, go off to great distance and to these corresponds a mean current of ie = 4qn to *which is due the action at a distance*. The total average current in the antenna is thus Im = 4Qn + 4qn and its intensity is no criterion for the performance. The electric efficiency of the antenna is q/Q+q and this is often a very small fraction.

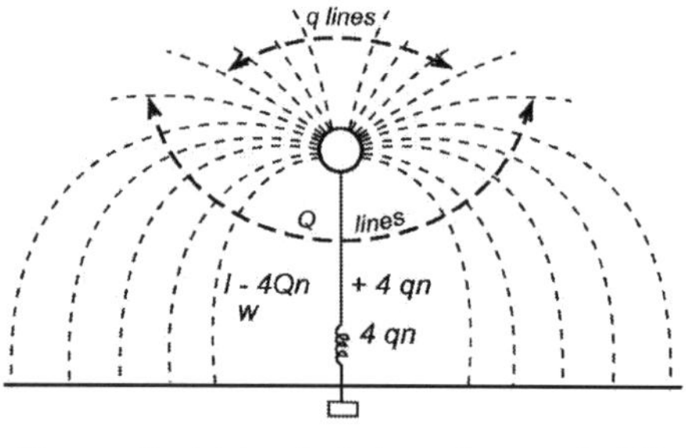

Fig. 14. Diagram Explaining the Relation Between the Effective and the Measured Current In the Antenna.

Dr. L. W. Austin and Mr. J. L. Hogan have made quantitative measurements which are valuable, but far from supporting the Hertz wave theory they are evidences in disproval of the same, as will be easily perceived by taking the above facts into consideration. Dr. Austin's researches are especially useful and instructive and I regret that I cannot agree with him on this subject. I do not think that if his receiver was affected by Hertz waves he could ever establish such relations as he has found, but he would be likely to reach these results if the

Hertz waves were in a large part eliminated. At great distance the space waves and the current waves are of equal energy, the former being merely an accompanying manifestation of the latter in accordance with the fundamental teachings of Maxwell.

It occurs to me here to ask the question—why have the Hertz waves, been reduced from the original frequencies to those I have advocated for my system, when in so doing the activity of the transmitting apparatus has been reduced a billion fold? I can invite any expert to perform an experiment such as is illustrated in Fig. 15, which shows the classical Hertz oscillator altho we may have in the Hertz oscillator an activity thousands of times greater, the effect on the receiver is not to be compared to that of the grounded circuit. This shows that *in the transmission from an airplane we are merely working thru a condenser*, the capacity of which is a function of a logarithmic ratio between the length of the conductor and the distance from the ground. The receiver is affected in exactly the same manner as from an ordinary transmitter, the only difference being that there is a certain modification of the action which can be predetermined from the electrical constants. It is not at all difficult to maintain communication between an airplane and a station on the ground, on the contrary, the feat is very easy.

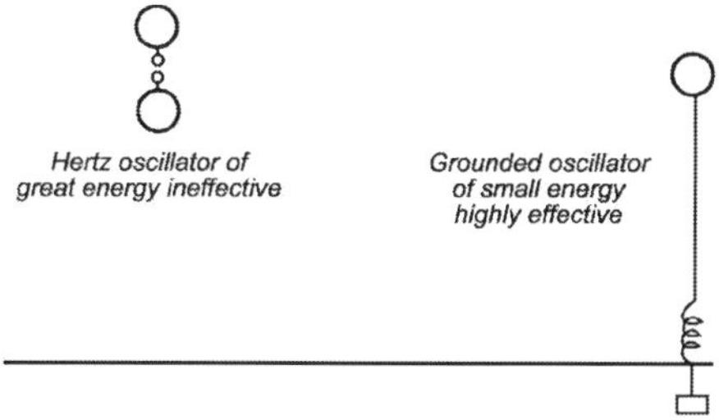

Fig. 15. Illustrating One of the General Evidences Against the Space Wave Transmission.

To mention another experiment in support of my view, I may refer to Fig. 16 in which two grounded circuits are shown excited by oscillations of the Hertzian order. It will be found that the antennas can be put out of parallelism without noticeable change in the action on the receiver, this proving that it is due to currents propagated thru the ground and not to space waves.

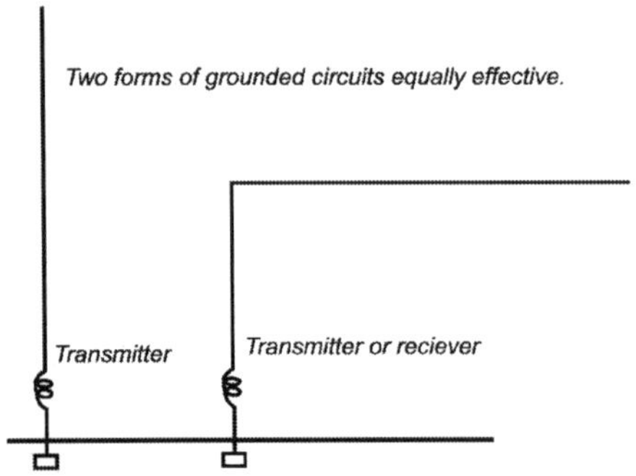

Fig. 16. Showing Unimportance of Relative Position of Transmitting and Receiving Antennae In Disproval of the Hertz-wave Theory.

Particularly significant are the results obtained in cases illustrated in Figures 17 and 18. In the former an obstacle is shown in the path of the waves but unless the receiver is within the effective *electrostatic* influence of the mountain range, the signals are not appreciably weakened by the presence of the latter, because the currents pass under it and excite the circuit in the same way as if it were attached to an energized wire. If, as in Fig. 18, a second range happens to be beyond the receiver, it could only strengthen the Hertz wave effect by reflection, but as a matter of fact it detracts greatly from the intensity of the received impulses because the electric niveau between the mountains is raised, as I have explained with my lightning protector in the EXPERIMENTER of February.

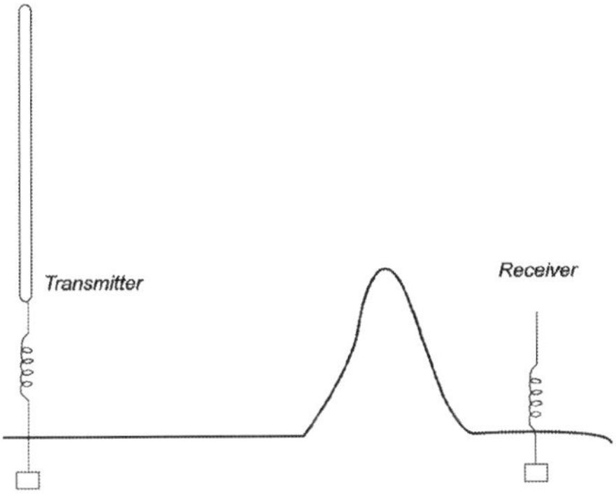

Fig. 17. Illustrating Influence of Obstacle In the Path of Transmission as Evidence Against the Hertz-wave Theory.

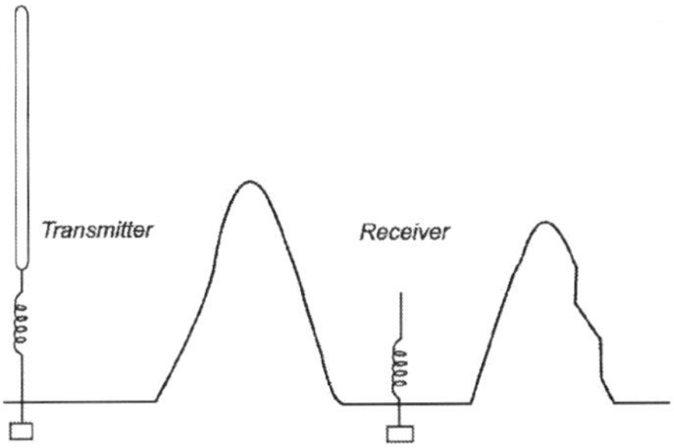

Fig. 18. Showing Effect of Two Hills as Further Proof Against the Hertz-wave Theory.

Again in Fig. 19 two transmitting circuits, one grounded directly and the other thru an air gap are shown. It is a common observation that the former is far more effective, which could not be the case with Hertz radiations. In a like manner if two grounded circuits are observed from day to day the effect is found to increase greatly with the dampness of the ground, and for the same reason also the transmission thru sea-water is more efficient.

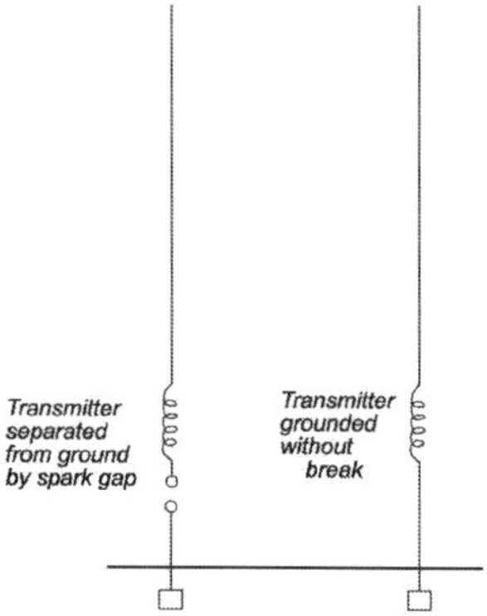

Fig. 19. Comparing the Actions of Two Forms of Transmitter as Bearing Out the Fallacy of the Hertz-wave Theory.

An illuminating experiment is indicated in Fig. 20 in which two grounded transmitters are shown, one with a large and the other with a small terminal capacity. Suppose that the latter be 1/10 of the former but that it is charged to 10 times the potential and let the frequency of the two circuits and therefore the currents in both antennas be exactly the same. The circuit with the smaller capacity will then have 10 times the energy of the other but the effects on the receiver will be in no wise proportionate.

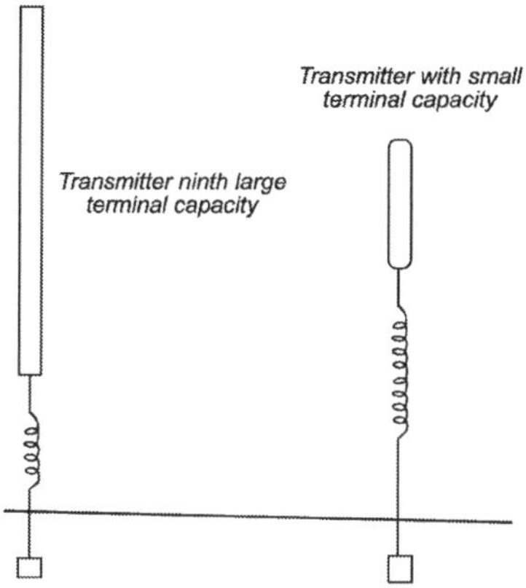

Fig. 20. Disproving the Hertz-wave Theory by Two Transmitters, One of Great and the Other of Small Energy.

The same conclusions will be reached by transmitting and receiving circuits with wires buried underground. In each case the actions carefully investigated will be found to be due to *earth currents*. Numerous other proofs might be cited which can be easily verified. So for example *oscillations of low frequency* are ever so much more effective in the transmission which is inconsistent with the prevailing idea. My observations in 1900 and the recent transmissions of signals to very great distances are another emphatic disproval.

The Hertz wave theory of wireless transmission may be kept up for a while, but I do not hesitate to say that in a short time it will be recognized as one of the most remarkable and inexplicable aberrations of the scientific mind which has ever been recorded in history.

Made in the USA
Monee, IL
04 November 2026

67352882R00017